DNA Healing Techniques

The How To Book of DNA Expansion and Rejuvenation

Robert V. Gerard, M.S.

with Todd Ovokaitys, M.D.
and Vianna McDaniel

Second Edition

Published by
Oughten House Foundation, Inc.
Livermore, California USA

DNA Healing Techniques
by Robert V. Gerard
Second Edition

Copyright © 1998 Robert V. Gerard
00 99 98   0 9 8 7 6 5 4 3 2
ISBN: 1-880666-77-4

All rights reserved. No part of this book may be reproduced, stored in a mechanical retrieval system, or transmitted in any form by electronic, video, laser, mechanical, photocopying, recording means or otherwise, in part or in whole, without written consent of the publisher.

The trademark affixed to DNA ACTIVATION™ has been used to avoid commercial and fraudulent use of the techniques and program as conceived and sponsored by the Oughten House Foundation, Inc. It is the Foundation's intent to do whatever is necessary to keep the DNA Healing Techniques simple, consistent, and unaltered. To keep commercialism out of the book, only a few insertions of the trademark will appear.

Published by
OUGHTEN HOUSE FOUNDATION, INC.
PO BOX 3134
LIVERMORE, CA 94551
PHONE (925) 447-2372 & FAX: (925) 798-5848
E-MAIL: dawne@value.net
E-MAIL: Robs1World@aol.com

SECOND EDITION: APRIL 1998

Disclaimers

This is an alternate healing practice, specifically, healing and rejuvenation of one's mental, emotional, and spiritual bodies. No claims are made regarding any medical or psychological condition or impairment. As always, in the event of any of the symptoms described herein, seek the advice of a physician. Please do not meditate while driving the car or operating machinery.

Working from the heart with the purest love and intent fosters the greatest benefits being given and received through these techniques. The most appropriate attitude with which to work is that this DNA healing information is a gift from Spirit.

Contents

Acknowledgments 4
DNA Project Mission Statement 5
A Perspective 6
Purpose of this book 9
Introduction 12
Part 1: DNA—Biological Master Molecules
by Dr. Todd Ovokaitys 15
Part 2: Overview of the DNA Healing and Rejuvenation
Techniques 25
Part 3: DNA Healing Techniques 37
    Technique #1: Activation 39
    Technique #2: Expansion and Rejuvenation 51
    Technique #3: Reading/Healing 57
Part 4: Transitioning after DNA Activation 65
    a. Reported Results, Symptoms, and Side Effects 65
    b. Meditation, Nutrition, and Exercise 69
Epilogue 71
Certified Training Programs 73
Discovering Your Destiny Programs 75
Suggested Reading 78
Audio Tapes 79
About the Authors 80
Monitoring Chart 81
DNA Activation Reference Guide 83
Testimonials 86
Personal Notes 88
Fund Raising and Donations 96

Acknowledgments

The authors wish to express their heartfelt gratitude to the following contributors who helped put this book together expeditiously:

To Kevin Shamel, for assisting Vianna with the initial manuscript

To Chris Hugh and all the fine folks at Nature's Path

To the staff at Gematria for their advice and input

To Dawne Fontaine, for her diligence and dedication in manuscript development

To George Stewart who allowed us to use the DNA illustrations on pages 16 and 18

To Tony Stubbs for his editorial detail and ongoing commentary

To the staff of Oughten House who volunteered their time to this project, and

To the members of Oughten House Foundation, Inc. who with their support made this publication happen.

A special thanks goes to all the Certified DNA practitioners and interns who have offered valuable experiential input, testimonials, and feedback. The information received from them has helped tweak and refine the techniques and simplify the presentation of this second edition.

With all due respect, I must also acknowledge Source and Spirit's intent behind the DNA Healing Techniques. It's best said this way: "The DNA Activations accelerate Spirit's relentless drive to lovingly resolve issues."

DNA Project Mission Statement

The mission of the DNA Project is to activate healers, teach the DNA Healing Techniques worldwide, and promote ongoing research into new ways of using the techniques. This endeavor also includes the dissemination of educational programs that complement the DNA healing process. Each program has been specifically designed to provide tools and techniques that encourage the healer's self-mastery and the discovery of personal destiny, and fosters the evolution of the Lightbody.

At the forefront of this mission is the recruitment of heart-driven healers and others dedicated to the mission. The people and the healings performed perpetuate the organization behind the mission.

A Perspective

Scientists and medical researchers have conquered what was deemed impossible fifty years ago. Yet as more is discovered, the opportunity of the unknown reveals itself even more. Our explorations have brought us to the outer limits of the universe proving that there are 100 billion suns within a typical galaxy and 200 billion galaxies within the observed universe. Fantastic? Yes!

Looking within our bodies is equally profound. According to ancient Hermetic Principles, the dynamics of cells are parallel to the dynamics of galaxies. The vast space that exists between the electrons and nucleus of each atom within various molecules is consistently being explored by quantum physics. Now we can look within and find that the most powerful electron microscope cannot reveal the magnificent organization of DNA within our own cells. What causes the particular DNA base molecules to sequence themselves in ways that contain the specific blueprints for our bodies? What causes aging and disease? Will we ever find out?

A worldwide initiative, the Human Genome Project, is exploring the genetic code, and has already cost more than was spent putting a man on the Moon. Research is still incomplete as thousands of researchers worldwide seek to better understand every detail of our DNA. Mapping out the chromosomes, which are made up of DNA, involves over 100,000 genes. Once identified, the "key" to life may be known. But let's go one step further and ask: "Who put the key there?" and "What intelligence is guiding the DNA to perform the most intricate and perfected tasks mankind has ever witnessed?"

The unique attributes about DNA and RNA center around their ability to replicate and communicate. Cell division is a marvelous process which deserves the label "miracle." Did you know that when a cell divides and replicates itself, the entire DNA also replicates itself? Conception itself is a miracle. Hold a newborn infant in your arms and see for yourself.

With regard to health and rejuvenation, it's our contention that substantial information has been brought forth suggesting that "mind over matter" begins when the DNA is instructed to do something. The best-selling author Dr. Deepak Chopra has written an entire section "DNA and Destiny" in his book *Ageless Body, Timeless Mind*, indicating that "awareness" is influenced by certain hormones within the DNA molecular process. He further states "The responsibility for changing this awareness lies with each individual." The highest form of intelligence is the ability to communicate. Communication requires us to be aware of our intent. And when we raise our intent to the highest levels of knowingness, we begin the process of creation.

In his newly released book, *Solarian Legacy*, Paul Von Ward discusses how consciousness plays a vital role in morphogenetic process, i.e., the development of the structure of an organism. He and other scientists now know that subtle energy fields exist throughout the known universe. Von Ward eloquently suggests that when "conscious intent" merges with these "subtle energies" creation results in both the stellar and atomic realms.

DNA appears to be complex, but it consists mainly of simple sugar and four base molecules. How do they know what to do? Your life depends on the way these molecules think, act, influence and replicate. Do we have a say in what they do?

The world and our lives are rapidly changing. Time seems to disappear right before our eyes. Predictions of the "end times" are all about us and we ask: "What is really happening?" and "Why?"

It is our contention that the era for mankind needs to take a quantum leap for itself. Our mental and physical health has become too dependent on others. We are faced with being lead and even misguided by what "they" say. It's always "them." *You* have a vital part in what you do with your health, when speaking your mind, and creating those things in life which bring you joy. The time is now! Take responsibility for your life—go within. Learn that self-mastery is your gift. You possess all the conscious intent and subtle energies to change your realties, including your DNA. Activate yourself, acknowledge your divine self. Become all you can be!

Purpose of this Book

This booklet describes techniques that contribute to individual and worldwide healing. If they resonate with you, then use them well. If they do not resonate, please pass the book along to a friend, or save it until a later time.

Our organizations wholeheartedly believe that this healing process is your birthright. As you develop this aspect of Life and restore your mind, body, and emotions to perfect health, you can then demonstrate the techniques and pass them on to others. Healing is our intent.

As a creator-healer, you will reap the benefits of perfect health and the joy of universal service. So, please remain open-minded and give these techniques a chance.

First, we suggest that you quickly read the booklet from beginning to end to get a feel for it. Then read it slowly and thoughtfully at least one more time. Then reflect on whether you desire to proceed. If you do, study each technique and learn the steps before applying them to others. All this could take you a few hours. Begin with the *Youth and Vitality Activation* technique and then the *DNA Expansion and Rejuvenation* technique. Next you might perform the *Reading/ Healing* technique. Again, practice a few times. Then honor your achievements and express gratitude for your newly learned healing talents.

The Oughten House Foundation, Inc. has and will continue to provide educational programs, support staff, and an interactive journal to assist you with these healing techniques.

However old these techniques may be, they are new to us. There have always been reports of miracles and mystical healings, but exactly how they occurred is unknown to most of us. As each day passes, more and more feedback will be received and reported to you. So please register your name with us, and we will keep you informed.

This book has been written in collaboration with the talents of Vianna McDaniel and those fine people at Nature's Path in Idaho Falls, Idaho; Robert V. Gerard, Dawne Fontaine, and Tony Stubbs of the Oughten House Foundation, Inc. in Livermore, California; and Dr. Todd Ovokaitys, M.D of Gematria Products, Inc. in Carlsbad, CA.

In order for the world to benefit from our discoveries, the information in this book must be learned, practiced, and taught. This is where you come in. Please share this with everyone, and help us spread the knowledge. We thank you.

In Love and Light

Self-Administering the DNA ACTIVATION™ Technique (First DNA Technique)

Many people have successfully "activated' themselves by using the first DNA ACTIVATION described. The process is straightforward, though, it is best to practice it a few times beforehand. There is a companion DNA ACTIVATION Tape that you can follow as some people prefer to listen. Use whichever method serves you best. Keep in mind, however, that the effect of the technique is sufficient for you to realize an extraordinary experience and even a spontaneous healing.

A few key points to consider:
- THE DNA ACTIVATION is a special event, it's truly a sacred signature of time and place
- Find a comfortable place with no interference
- Learn how to get into a very relaxed mental state, which we refer to as a "theta brain wave" frequency
- Trust yourself
- Let the DNA ACTIVATION process flow through you
- Avoid putting expectations on it, that's a mental process; the technique is "heart driven"
- It's okay to repeat the technique

- Don't be surprised if you see colors, sights, or sense energy moving through you
- Enjoy it
- Give us a call or write to us, we'd love to hear from you

These and most healings can be done remotely, once comfortable with the techniques. However, we do ask that whenever someone is activated that they purchase this book for reference sake. For follow-up purposes and support for those ACTIVATED, the Oughten House Foundation can periodically send information and updates as to the many benefits and progress of the DNA Healing Project. So please send the Foundation the mailing address of those ACTIVATED.

Note: In this book, wherever we use the term "God," we refer to the Creator of All That Is. Please substitute whatever term you are comfortable with or prefer to address your deity. We also use the term "client" to refer to the person on whom the healing is done.

Introduction

The three DNA Healing Techniques are powerful yet simple ways to explore inner and psychic space. Possibly for the first time in your life, you will witness a spontaneous healing, see a pulsar light in your mind's eye, and feel larger than the room you're sitting in. Experiencing the DNA Activations is the first step toward total integration of body, mind, and soul—a moment you'll joyfully remember for the rest of your life. As you become more familiar with the techniques, you will learn how to psychically enter someone's body to effect healing. You will become familiar with the most intricate details of the human body and discover the vastness of the inner world. Imagine venturing inside the cells of the human body, talking to them, and commanding them to function correctly. Imagine transforming someone's pain and dis-ease into Love. Imagine the power of healing and deeply understanding the way things really work. Most importantly, *imagine*.

Our known universe consists of trillions of stars and planets. The space between all this matter is not empty, but is filled with many different types of energy. There are 10 trillion cells in the human body, and through vast and complex processes, they organize and communicate with one another. Each cell is conscious and aware of itself and the others, and has a specific function to perform in concert with all the other cells.

Billions of dollars are spent annually on genetic research into cells and their DNA. The core issue, as we understand it, is more subtle—learning how the cells and their DNA *communicate!* Once that is identified, the mysteries of life can unfold as we listen to what is being said, or maybe, the DNA will listen to us.

Part 1 of this booklet explains the workings of DNA. It provides factual and detailed information about the complexities and wonders of the blueprints of life. Carefully written by our medical advisor, Dr. Todd Ovokaitys, it explores the intricacies of DNA and some recent discoveries in the field of DNA research.

Part 2 provides a general overview and orientation to the three DNA Healing Techniques, plus some preparatory instructions and excellent visualization techniques. The phrase "mind over matter" comes to life in this section as you see how the miracle of Life is based on the Divine Blueprint, and how these techniques will empower mankind. However, it cannot be emphasized enough that success with these techniques requires the *purest intent* and *love from the heart*.

Part 3 provides step-by-step instructions for the techniques. You will learn the real power behind conscious intent, commanding communications, trust, and imagery. The process is explained along with sufficient guidance and detailed scripts to apply the techniques.

This booklet is a journey to better health, healing, and sharing. It will take you to the edge of your imagination, unlock doors of limitations, and propel your creativity to dimensions once only dreamed of. We encourage you to read this little book, test it, and see for yourself whether you are ready to be a co-creator of Life—in the image and likeness of the Creator (however you perceive the beauty in that).

Part 1: DNA—Biological Master Molecule

[By Dr. Todd Ovokaitys, medical doctor pioneering in laser DNA research]

The purpose of this chapter is to inform and enliven your perceptions of the workings of DNA. The fundamental power of DNA is the outplay of a simple coding system into rich varieties of life expressions. DNA will be defined, its structure described, and the delicate architecture of its operations revealed. Knowing how DNA works is an introduction to the most intimate aspects of your biology.

DNA gives the instructions for form, cell by cell, to every living thing from microscopic "wrigglers" to hundred ton whales. DNA tells a cell what it has been, what it will continue to be, and what it will become. The timeline encoded in DNA defines cycles of growth, repair, replication, and eventual dissolution for all the cells of a living organism.

DNA regulates many levels of information. DNA carries the ancestral imprints of all preceding generations to the present. DNA operates the biological clocks that count the days in each cell of your tissue and sets the life span. To go beyond the usual limits of longevity will require learning how to reset the time clocks controlled by DNA.

The initials of DNA stand for deoxyribonucleic acid, pronounced dee-ox-ee-rye-bo-noo-clay-ic acid. Dissolved in the cell, DNA has the properties of a weak acid, many times weaker than the battery acid that powers your car. "Nucleic" refers to the fact that DNA resides in the cell's nucleus within a specialized nuclear membrane. (See Figure 1.)

The basic structure of DNA is two very long chains wound around each other, one going east the other going west. The popular image is that of an outstretched spiral staircase. The two side supports are the backbone of the structure and maintain its integrity of form. The rungs or treads between the supports carry the specific information of DNA.

The "deoxyribo" part of DNA refers to the name of a simple sugar molecule called deoxyribose. This sugar is even simpler than the glucose that gives Gatorade its punch. Deox-

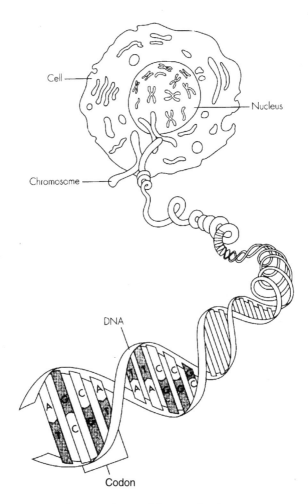

Figure 1: Cells and Chromosomes

yribose is part of the backbone of the DNA spiral stairway, one molecule of deoxyribose every step of the way on both sides of the DNA structure. A single phosphate molecule connects consequent dioxyribose units to complete the DNA spiral staircase rails.

The staircase treads holding the actual information are made up of only four different molecules called bases. The names of these bases are adenine, cytosine, guanine, and thymine. The shorthand abbreviation for these molecules is simply the first letter of the chemical name: adenine-A, cytosine-C, guanine-G, and thymine-T.

Each rung of the stairway is made up of a pair of bases. Each side support unit of the staircase projects one base into the center of the spiral. The four bases pair up in a very specific way. Adenine always pairs with thymine, and cytosine only links with guanine. Therefore, whenever you see a T on a DNA rung you know that an A will be across from it, and a G will be across from a C. For example, if the sequence on one chain were A-T-A-G-C-G, then its partner chain would be T-A-T-C-G-C.

The precise complementing of the DNA base pairs sustains the integrity of the code from generation to generation. When a DNA molecule replicates into two copies of itself it follows a very simple process.

The two chains split apart in the middle. The pattern on both halves attracts its complementary pattern to itself to form two new complete staircases. So, chain 1 separates from chain 2; chain 1 builds onto itself a new copy of chain 2; and chain 2 constructs onto itself a new copy of chain 1. This process occurs with such accuracy that many DNA sequence patterns have remained consistent for millions of years. In the individual organism, duplication of DNA is required for cells to divide and provide each new cell with its own copy of the full DNA blueprint. DNA directs its own and overall cellular replication to renew and replace cells as needed to sustain tissues in good repair (see Figure 2).

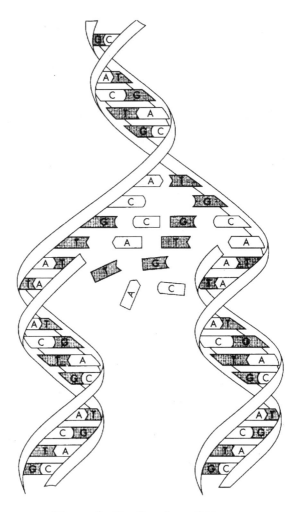

Figure 2: Replication of Genes

The DNA code is read in a specific direction. Each double helix of DNA consists of two very elongated molecular chains that wrap around each other. The information in one chain is read in the opposite direction from the other chain. Sequence patterns in the chains instruct the cell chemistry as to which parts of the DNA should be read and in which direction to get the code.

Once a sequence in DNA is activated for expression, the mechanism of decoding is precise. Each sequence of 3 bases, known as a *codon*, delivers a specific instruction to the cell chemistry. Since there are 4 different bases, the number of combinations of 3 bases in sequence is 4x4x4, or 64. Thus, there are exactly 64 different codons that regulate cell chemistry and function.

The most central operation of codons is to direct the manufacture of proteins. A protein is a complex molecule made up of a string of simpler building blocks called amino acids. A protein can be structural like collagen or elastin that supports the form of a tissue. Proteins can also be functional such as muscles that contract, or enzymes that catalyze the many chemical reactions that preserve the life of the cell. Codon commands include: start a new protein chain, add a specific amino acid to the chain, or stop the chain growth at this exact position.

Much as a shop that makes custom nuts and bolts, DNA has a procedure for making well-crafted proteins:

1. The part of the DNA that is to be read has its sequence information copied onto a molecule called RNA, or ribonucleic acid.

2. This RNA departs the nucleus and goes to the liquid part of the cell called the cytoplasm. (The process of DNA copying its sequence pattern to the RNA is known as transcription, much as a secretary accurately transcribes dictation. Because this form of RNA delivers information from the nucleus to the rest of the cell, it is called messenger RNA, or mRNA.)

3. RNA then directs the process of translation of the DNA sequence into the protein encoded in the pattern.

4. A start codon initiates the process with the amino acid methionine.
5. The subsequent codons tell the cell machinery which of the 20 amino acids to add next to the growing protein chain. This is the elongation phase of protein synthesis.
6. When the sequence of adding amino acids is complete, the termination codon tells the cell "job well done, on to the next product."

Thus DNA is the blueprint for the structural and functional proteins that condition and regulate life processes. A DNA *sequence* that encodes for a protein is known as a *gene*. The information content of human DNA's double helix is so enormous that approximately 100,000 genes reside within it.

The term *genome* refers to the entire genetic code for an organism. The nucleus of each cell of the body contains the full genome, the complete genetic code for the whole body. Thus, each cell contains the full blueprint needed to recreate the entire organism.

In simple life forms like bacteria, a single chain of DNA is adequate to encode all the needs of the organism. More complex forms require more information and greater lengths of DNA than can be housed on one strand. When the DNA is divided into multiple chains in a cell, as in human cells, each chain of DNA is called a *chromosome*. The full complement of human DNA is divided into 46 chromosomes, each one containing characteristic genes that distinguish one chromosome from another.

When researchers first stained cells with dyes to see them under a microscope, chromosomes were seen to take up color avidly. The word chromosome derives from the Greek word "chromo" for color and "soma" for body. Thus chromosomes are "intensely colored bodies in the cell."

The 46 human chromosomes form 23 pairs. One member of each pair comes from the mother and one from the father.

Each egg and each sperm contain exactly 23 chromosomes so that when egg and sperm unite, the full count of 46 chromosomes is realized. The father's chromosomes determine sex. There are two types of sex chromosomes, X and Y. The pattern of combining X and Y chromosomes directs whether the baby will be born boy or girl. Each egg has one X chromosome. Each sperm has a 50:50 chance of bearing an X or Y chromosome. If the father's sperm has an X chromosome when it combines with the egg, the baby will be a girl (XX makeup).

If a sperm with a Y chromosome wins the swimming contest to the egg (fertilization), the result will be a boy (XY makeup). Therefore, it is the coin toss of paternal sperm makeup that decides the sex of offspring. Since sperm with Y chromosomes are lighter than those with X chromosomes, they can swim a little faster, so, over a large population, a few more boys tend to be born than girls. Thus, even at the sperm level, men reveal their impulse to compete.

The other 22 pairs of chromosomes are called somatic chromosomes or those of the soma or body in general. These are numbered 1 through 22, and each has features which clearly distinguish it from all the others.

One of each number comes from the mother and one of each from the father. The full human genome thus has 22 pairs of somatic chromosomes and one pair of sex chromosomes. These 46 chromosomes act in harmony and generate the entire body from a fertilized ovum. The genome information then directs further development and the sustaining of all life processes.

The full complement of human DNA is a huge database approximately 3 billion base pairs long. Many genes have been mapped onto the specific chromosomes on which they reside. By their products, other genes have been identified, but their exact chromosomal address has not yet been determined.

The Human Genome Organization, known as HUGO is attempting to decode the base sequencing of all human DNA. This massive analytical and data gathering task has been taken on by a multinational team of scientists, each with their bit of DNA to sequence. It is estimated that in as little as 10 years, the entire human genome will be completely mapped.

If the DNA in all 46 human chromosomes were stretched out end to end, the length would be about 6 feet. Within the nucleus of a cell this expanse of DNA folds into a space of a few thousandths of a millimeter. How is this possible?

In the basic double helix of DNA, there is one complete turn of the spiral every 10 base pairs. The average strand length has about 7 million full turns. The DNA strand then forms a coil within a coil containing eighty base pairs per loop. This structure then coils again and again up to 5-7 levels of supercoiling overall. It is through the "coil within coil" patterning that the full length of DNA can be packed into a tiny nucleus.

In order for a gene to be transcribed, the supercoil must unfold to expose the desired region of DNA. The unfolded DNA must then "unzip" down the middle to form a template to build the messenger RNA molecule that will direct construction of the intended protein product.

The DNA then rezips, recoils, and awaits another call to action. Unlike most clothing zippers, the action is fast, the mating is clean, and there are no awkward snaps.

It is very important that the ends of chromosomes avoid the tendency to tangle when all 46 chromosomes are dividing. To prevent an awful balled up mess, the ends of chromosomes have particular sequences that repeat many times to maintain chromosome integrity during division. These vital regions are known as *telomeres* ("telos" a Greek root for end and "mere" a designation for a chromosome region). Telomeres are, therefore, the chromosome's end-caps.

Telomeres also appear to have a biological clock function, and as such, have been the object of recent intense scientific study. When normal cells divide, the telomere regions of chromosomes tend to chip off, and over many cycles of division, the telomeres may be whittled away completely. In this case, the cell loses the ability to divide. The wearing down process of telomeres has been identified as a major reason why human tissue lives only so long. The telomeres may be one of the most important, if not the main, determinant of human life span.

In contrast, a cancer cell produces an enzyme called telomerase. This enzyme restores the telomeres to their full length. Cancer cells may divide endlessly, unlike normal cells, which usually stop replicating after a certain number of divisions. A new approach for treating cancer may be the use of agents that inactivate telomerase, so that cancer cells, too, may obey the signal to stop dividing.

On the flip side, a method which turns on telomerase in normal cells, could retard or even reverse the aging of healthy tissue. Resetting the biological clock by restoring the full length of telomeres is the first true candidate for extending human longevity, leading even to biological immortality.

From the humble origins of noting that chromosomes attract colored dyes, genetic science has come a long way. Within a decade, the detailed sequence of the full genome may be charted. This information and telomere biology may propel mankind to new levels of quality and duration of life.

Part 2: Overview of the DNA Healing and Rejuvenation Techniques

Part 2 provides a brief orientation to the DNA ACTIVATION and Healing process, brain states, tones, colors, latent DNA structures, expansion and rejuvenation, and explains why communication and trust are such important aspects of healing.

Part 3 presents descriptions, usage, and instructions for the three techniques presented in this book:
1. Youth and Vitality Chromosome Activation, or *DNA Activation*
2. DNA Expansion and Rejuvenation or *Expansion/ Rejuvenation*
3. The Whole Body Diagnostic Reading/Healing, or *Reading/Healing*

Overview: The DNA ACTIVATION and Healing Process

First and foremost, the DNA ACTIVATION is a process. It has nothing to do with an individual's religion or beliefs. It's as easy as taking a brisk walk or cooking an omelet. Exercise and food are vital essentials to physical health, and so is the DNA ACTIVATION. The process only takes a few minutes, but the results will last a life-time. I consider it a sacred event; sacred in the sense that it will be remembered as a very special and simple gift received from the Creator. And that's the only problem I can warn you about: it's simple and people hesitate to realize its potential and eventually forget about it. It's how we easily forget the importance of baptism, confirmation, and bahmitzvah.

The DNA ACTIVATION process is straightforward. It can be easily conceptualized as the integration of the loving God within you, that is, your True or Higher-Self, your Deity, and your body consciousness that belongs to the Planet, which I loosely term "your DNA's Intellect." This is a new trinity, if you will, a new relationship wherein your total true-self, your Deity, and the DNA within your body are in synchronization. They are now one. For most, this experience is easily realized when a person receives the DNA ACTIVATION™. In many occasions, a spontaneous healing, either physical or emotional, is experienced.

The new trinity diagram looks like this:

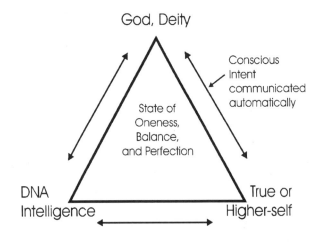

When God consciousness and DNA consciousness talk, destiny is revealed. Once the DNA's Intellect has been activated, it now becomes one with the relationship of the new trinity, no longer existing as having dominion over it. This in effect is a "Great Shift" in consciousness. The body now has equal domain of its own existence. In all reality, however, we do have free will and choice, and can easily and rationally over-ride this simplistic phenomenon and maintain our separation from God, the Planet, and remain in our burdensome duality. We can use our free will to be "one with God" and end separation as everything other than love is duality. In this new trinity, the intent for Source, DNA Intellect, and your Higher-Self automatically communicate. What is revealed to you is intuitively provided as long as we keep an open ear. Wisdom becomes the forefront of your decision-making as information about your health, your emotions, and destiny are subtly brought forth. The results obtained thus far in the DNA Healing Project have consistently demonstrated that these trends exist.

Brain States

Working and altering brain waves have become standard procedures in biofeedback, research medicine and psychology. These brainwaves can either be all working at once or separately. In sports, athletes easily demonstrates this phenomenon. A figure skater or skier for instance must be fully active (beta) yet be focused (alpha) and emotionally free and spontaneous (theta). In prayer or meditation, a person would be in *alpha* and move in and out of *theta*. Brain states can be changed in various ways: breathing, mental or emotional experience, sensory patterning (light, sound), movement or body stance, chemically or nutritionally. There are plenty of books and research papers available on brainwaves, their effect on consciousness, and how it all works. Below are the four brainwave categories:

- *Beta*: 15 to 38 cycles per second. Beta is characterized by a high state of alertness and active consciousness. This is our normal waking state.
- *Alpha*: 8 to 14 cycles per second, related to relaxed wakefulness and light meditative states. A person is aware of his or her body, accelerated learning happens due to less stress around learning.

- *Theta*: 4 to 7 cycles per second, associated with trance-like mental, visual or imaginative states, deep meditation, light sleep mental states, and most aspects of creativity. In Theta there can be visual images, and possibly little or no sense of the physical body. A person may experience his body as an energy field or body parts floating but not connected. The experience one has during *theta* activity seems to be the most prolific. Theta is the ideal state for some types of self-programming, accelerated learning, and self-healing.
- *Delta*: 0.5 to 3 cycles per second, usually related to deep sleep. In the lowest levels of Delta, there are no mental images and no awareness of the physical body. Some mediators experience deep a meditative state: asleep yet fully conscious. Reports indicate being aware of self, as still awareness, not as a person or a body.

Let's try to feel what moving into and out of the "theta" frequency feels like. Become very active or mentally occupied for a few minutes. This usually brings you into *beta*. Sit down, relax, and take a couple of deep breaths. You may sense moving into an *alpha* state. Take a couple more deep breaths and let yourself drift a bit. When your mind is free of thinking of things and your thoughts float, then you may be experiencing *theta*. If you fell asleep, you went into *delta*.

Here's a good exercise to try while sitting down. First take a few deep belly breaths to get you into an alpha state. Picture a shiny pulsating star of light in your chest. Now let it rise. Use your imagination to help. Let it rise through your throat, through your head, and up over yourself about 3 feet. Let it stay above you for a moment or two, then slowly bring it back, reversing the process. At this point, you're most likely in *theta*. You should feel quite still, focused, and mentally unencumbered. This is a good *theta* warm-up exercise.

The DNA Healing Techniques use *theta*, or 4 to 7 cycles per second. It is a very slow state of activity, and normally occurs during deep meditation, hypnosis, or light sleep. You will learn to slow your brain to this state, and it will allow you to project your consciousness from your physical body into that of the client's. Interestingly, your client will also slip into a theta state. For many of you, this will simply be remembering an aspect of your being that you're not always aware of but that's completely natural to you.

You will find that the DNA Healing Techniques expand the more you use them. As you learn the basics, you'll discover more and more. As each day passes, we are making important discoveries about new ways of using the techniques to better the human race, and you, too, will learn more each time you use them.

Tones

Most of us enjoy our music of choice. Music can make us happy, make us dance, and even put us into an emotional state. In movie theaters, it's the music that brings out the excitement, tense drama, and romance. Turn off the music and a lot will be missing and you'll be amazed of the role that sound plays in our daily lives. Music is a tone. It resonates through you like no other substance. Tones are frequencies that carry messages.

The DNA communicates by *tones*. Tones are also how the intent of the human mind is connected or relayed to each DNA strand. The intent of the DNA strands is then communicated via tones to all other DNA strands instantly. This alters the chromosomes which in turn alters the cell. The corollary also exists: the DNA uses the tone to communicate its intent to the healer.

Cells communicate with each other, and the DNA Healing Techniques enhance cellular communication, not only within the client's body, but also between healer and client. This is an important factor when dealing with DNA changes. Once the healer's DNA has been changed, sometimes just physical contact will start a client along the process of cellular reconstruction. Imagine this process happening on a global level.

Latent DNA Structures

In addition to each chromosome's current two-strand double-helix of DNA, there are 10 strands (five pairs) that are "virtual," or not fully physically manifested, and need to be activated into our third-dimensional physical framework. Do these mysterious virtual DNA pairs hold the secrets of a healthier human body and race?

These ten virtual strands are actually a *cosmic link* or the universal life force residing in each cell. They connect spirit with the human genetic coding. Each strand is profound in its infinite capacity to enhance human life to that of "light-body" and may have universal functions beyond human limitations. There exists some kind of cellular language that connects consciousness in the cell with that of the entire body. Many studies and experiments by scientists in quantum physics, medicine, psychology, and metaphysics reinforce the view that somehow, a cellular language enables each cell to resonate within and beyond its intention and purpose.

We have named the *virtual* strands as follows:

- *Communication*: The command center to communicate and resonate with all existing life forms within and beyond the cell. This is where the conscious intent of the mind and body respectively exchange information, interact via tone, and where the intent of the soul is recorded in the physical body.

- *Perfection*: The never ending quest for health, perfect cellular replication, healing, and expansion of the light body. These are the strands that confront incongruities within the DNA itself and the immune system.

They receive and hear the command to heal and return to balance.

- *E-Motion:* Cellular identity and resonance, initiator of cellular ambiance and expansion, ultimate expression of happiness, joy, and bliss. Energy activation for love, truth, and trust. These strands of DNA process all psychological and emotional commands and memory. The assistance to identify and release emotional blocks and life traumas comes from these strands.

- *Creativity:* These strands co-create the "cultural" aspects of life. Caring aspects of conscious manifestation, intuitive surges from higher sources as well as inner cellular wisdom and abundance are here. It is the process of expressing the Soul and manifesting the Divine Blueprint. We heal our illusions and learn to be, not just survive.

- *Immortality*: Spontaneous and timeless living in the moment; eternity, and affinity with the Creator and Christ Consciousness. The gold and silver strands—Threads of Everlasting Life—stand for ultimate balance between all things, and assures that the relationship between the Creator, Man and the Planet remain in perfect harmony and oneness. We go into our own totality, our ownership of ourselves ... ultimately, Who We Are!

DNA structures and processes can be altered or damaged for a number of reasons such as toxins and radiation. As a result, when the cells replicate (reproduce), abnormalities can occur, or the cell itself may malfunction. DNA Healing Techniques can repair the damage and expand the DNA from the current two-strands to multiple strands.

The combined ability of the DNA strands is magnificent. They operate at the creative-subconscious level and are totally dependent on the *free choice* or *free will* of the individual. They are obedient beyond imagination, for the process of free will and choice are truly gifts from God. What makes them spectacular is that they operate within their own functionality. With total conscious creative action, they will instantaneously execute the "will" of the individual, blissfully performing every operation to perfection, with unlimited scope.

Colors

Energy is frequency and colors are their expression. The color spectrum offers a vast array of information and healing energy, from delicate hues to deep primary color bases. In most cases, individuals see energy wave patterns and often symbols. These are usually displayed in a variety of colors and combination of colors. For instance, I see and use a translucent blue healing tube about an inch thick for cleansing emotions and bodily pain. As I pass this translucent blue healing tube over and through the area of discomfort, I notice brilliant silver/white sparkles ignite as I move the tube. Each sparkle is a healing of some sort. As simple as this may seem, it's effective. Often I see brilliant green healing energy wrap itself around the organ or inflected area of discomfort.

There are many good books to read about the healing powers of color, providing valuable insight into the subtle healing gifts that are available to us. Colors play an important role in the DNA Healing Techniques. Certain colors are associated with the virtual DNA strands we have activated. (Much thought went into the question of this topic because of the myriad of variations associated with the healing energies of color. We did not want to set forth a rigid formula, but people inquired and thus we added this section into the book.)

As the user and practitioner of the DNA Healing Techniques gain in experience, so will their ability to see and use healing color energies. We advocate using healing color energies as they manifest in the healing session or ACTIVATION. To help better understand the virtual DNA strands, the following chart is provided, but do bear in mind that the colors can vary from person to person:

Communications	Blue
Perfection	Violet
E-Motion	Pink
Creativity	Peach
Immortality	Gold & Silver

Initial Preparation

First and foremost, it is extremely important that the DNA Healing Techniques be used with the **intent of love from the heart**. Before learning them, become comfortable with visualizations and using your imagination. The Pulsar Star exercise and Soul Bubble visualizations (see below) are excellent tools to move into the *theta* state. The more you practice, the quicker the results.

Once comfortable that you can attain *theta*, move on to any of the techniques. It is important to remember that each of the three techniques uses virtually the same starting point, with only minor variations in words.

"Pulsar Star" Exercise - A mini theta exercise

To get a feel for theta, try this simple "Pulsar Star" exercise. Begin by taking a couple of deep belly breaths. Imagine a beautiful star inside of your chest. Feel your energy like a pulsar at your heart. Move this star-energy slowly upwards through your throat, to the top of your head. Keep the vision. This equates to the Alpha brain state. Continue moving this star-energy up out of your crown chakra to about 3 feet over your head. Move the pulsar energy up to 6 feet over your head and hold it there for a couple of seconds. This is theta brain state. Enjoy the moment. Slowly begin to bring this star-energy back into your crown and into your heart. As you allow the star to drop, feel your body assimilate a theta vibration as well.

The Soul Bubble Visualization

Visualization/imagination is ironically one of the most powerful tools our species has, and one of the least appreciated. Begin with several deep "belly breaths," and move your point of awareness into your belly. This puts you into an *alpha* state. Then use your imagination to visualize your brain waves slowing down even more into *theta*—the creative state.

Next, try a visualization exercise called "The Soul Bubble," consisting of an alpha visualization which relaxes you, and then puts you in a theta state as your imagination creates a beautiful space in which you can actually feel your soul and consciousness.

This can be done individually as well as in a group. It's best, however, to have someone lead the group. Choose a room that is quiet or an area that you are comfortable in. Standing up may serve you best. Close your eyes and take several deep breaths to the bottom of your belly. Stand comfortably with your hands at your sides. Now you will build a translucent "soul bubble" around you.

About eight feet in front of you, imagine a translucent shield separating you from the outer world. This is the front of your soul bubble. Then imagine the shield on your left side, also about eight feet out, on your right side, and behind you. Do this eight feet above your head.

Now take a deep belly breath and imagine the translucent soul bubble eight feet below your feet. Sense the shields joining and forming a complete bubble around you.

As you begin to rise and become the nucleus of the soul bubble, you can actually feel your consciousness shift. You are rising into an altered state of higher consciousness. Remember this feeling, for it is the "real" you. You are centered. It's the you that can enjoy bliss. Feel yourself floating within the soul bubble in a perfect state of oneness. Remember this feeling always; it activates peace within you. This is your sacred space.

Meditation

Meditating provides an excellent vehicle to relax and go within. Too much information has been revealed substantiating the benefits of meditation for us to ignore. As a way to increase the experience of the DNA ACTIVATION™ and subsequent healings, I highly recommend some form of meditation. Meditating twice daily and doing a DNA Healing on yourself can improve your life substantially. I have met thousands of people who have improved their health, emotions, and their outlook on life as a result of their meditations. Examples of the DNA Meditations used in groups are given in Part 4 of this book.

Now that we have a clearer understanding of the mechanics of DNA, chromosomes, communication and trust, let's learn the techniques.

Part 3: The DNA Healing Techniques

Remember, you can use these techniques on yourself and on others. They are simple enough to do and don't be afraid of making mistakes. The worst scenario is that nothing happens, but I doubt that. Learn the three basic steps, practice, or use the companion DNA ACTIVATION tape.

Each technique has three major steps:
1. Divine Orientation (common to all techniques except for a few words).
2. The specific technique (Activation, Expansion/ Rejuvenation, or Reading/Healing).
3. Ending the Session (common to all techniques).

In addition, within the first and second major steps, there are secondary steps. These require specific actions to be performed depending on the process sought, for example: (a) Command: "Show Me...," and (b) state a Specific Process....

Details for all three steps are given in the Activation Technique. Because steps 1 and 3 are so similar for the other two techniques, they are not repeated in the descriptions of those techniques.

> *In all applications, these techniques must be used with the intent of love from the heart. This is of paramount importance, otherwise, nothing will be manifested. The secret is love, trust, and respect.*

Asking Permission

Ask the client's permission to do any form of healing work. This puts both the client and the practitioner in a state of balance and openness. Simply state: "_____ [client's name], do I have permission to do a healing with you at this time?" We also recommend that you record the session and give a copy of the tape to the client for reference. Activation sessions generally take about 30 minutes.

Technique #1: Activation of the Youth and Vitality Chromosomes
[aka DNA ACTIVATION]

Two archetypal chromosomes are primarily responsible for activating the 12-Strand DNA to its full potential. They are:

- The "Youth" chromosome is also referred to as the "Chronus" chromosome named after the son of the Greek mythological god Zeus. Chronus is the chromosome that counts every second, minute, and hour of your life. It's the timekeeper. This MUST be activated first.

- The "Vitality" chromosome, the most mysterious in nature, carries the key which unlocks the door permitting the dormant DNA to be activated to its full potential.

Step 1: Divine Orientation (common to all techniques except for a few words)

This is the starting point for all three techniques, and is virtually the same, except for a slight change in wording. There are six three steps in the Divine Orientation, and each step is meaningful and crucial.

a. Raise your consciousness:

- It's best for you and the client to sit facing each other. Place your palms under the client's palms. This creates an electrical circuit, and allows your cells to begin talking.

- Become centered in yourself, take a few deep belly breaths, pull all of your energy inward, focusing it in your heart. Balance your energy. Feel at peace. This puts you into an *alpha* state.

- Let your consciousness rise from your heart (the heart chakra) out of the top of your head (the crown chakra), going approximately three feet above your head. This takes your mind deeper into alpha. From this point, you can imagine yourself as a ball of light. Picture a spiritual or etheric version of yourself floating out the top of your head. This is a great feeling.

  By rising from your head and making the following command, the universe recognizes you. You will automatically be in truth, so that what you imagine will be real. You are not making this up.

b. CALL IN GOD, [or whatever term you prefer to address your deity]. Once you do this, you go into a *theta* state. This concludes the Divine Orientation.

Step 2: Specific Activation Technique

a. "I COMMAND _____" prefix (*state what you desire be done*). This calls forth the universe's powers of divine wisdom and manifestation to perform a specific action.

- The word "COMMAND" is important. This is the God-in-you speaking. This is your higher-Self claiming its magnificent power. You are not telling God to do anything, but you are calling on the powers of creation, the Universal Laws of Life, to serve you and the client.

- Remember, the "I COMMAND _____" prefix statement is the pivotal declaration of intent—to perform an Activation, a Reading/Healing, or a specific action. Be clear on your intent! As you make this command, you move your *etheric* self over to the top of the client's head, and by the time you finish the command, you should be entering the client's body through the top of the crown chakra. Remember, if you replace the "Command" word with an "ask or beg" intent, the activation may not work. It's important to get used to the intent and feelings behind the use of the "Command" word.

b. Use the following "I Command" for the First DNA ACTIVATION: "I COMMAND AN ACTIVATION OF THE YOUTH AND VITALITY CHROMOSOMES BE PERFORMED ON ......"

c. State the client's full name, date, time, and location where the activation takes place. In narrative form, it flows like this for the first technique. In your mind or state out loud, make the following Command, "*GOD*, I COMMAND AN ACTIVATION OF THE YOUTH AND VITALITY CHROMOSOMES BE PERFORMED ON, _____ (person's full name), ON THIS DAY (the whole date), _____ AT THE TIME (morning, evening, or a.m. or p.m.) at _____ (location)."

The preceding wording triggers an activation for the client. You would use slightly different wording for Expansion/Rejuvenation and Reading/Healing.

You have to hold the thought in the *theta* state for only TWO seconds, then you can release it. It does not matter if you doubt yourself later. It does not matter if the client believes you. You are bypassing the client's belief system and speaking directly with the cells.

d. In this *theta* state, enter the client's body through the top of the head (crown chakra) and into the pineal gland. Be very attentive to the inner *voice* of your highest consciousness.

e. There is a central cell in every person, perhaps the first cell formed in conception, located in the *pineal gland*. Upon entering the pineal gland, command: "SHOW ME THE GREAT CENTRAL CELL IN THIS PERSON," and you will be shown. Use your imagination to help the manifestation of the cell's image.

f. Enter the cell. Command: "SHOW ME THE CHROMOSOMES." Cells are beautiful on the inside and much more spacious than you would think. You are now smaller than a single cell, and you will see light flying around you as various parts of the cell perform their functions. In the center of all this activity, you will find the nucleus. Enter the nucleus of the central cell, and you will find chromosomes. They seem to be suspended, wriggling around, holding the patterns of human life within.

g. Command: "SHOW ME THE YOUTH AND VITALITY CHROMOSOMES." You will be shown the chromosomes or some form of symbolic representation. Pay close attention to your intuition and higher-self's voice. It is when you sense or know that these two primary chromosomes exist that they have communicated to you. You are now bonded in oneness and united in each other's intent.

h. Then command: "SHOW ME THEIR DNA." You will then see the original or archetypal biological 2-strand DNA helix uncoil.

i. You may have the client repeat the following commands to reinforce their own commitment to this process.

j. Then say: "I COMMAND ACTIVATION OF THE YOUTH AND VITALITY CHROMOSOMES IN ME NOW!" Once this command is given, the consciousness of the Youth and Vitality chromosomes will initiate a call to the DNA within itself, and begin to repair and heal.

At this point, transformation of the 10 virtual DNA strands (Communications, Perfection, E-Motion, Creativity, and Immortality) will be initiated and they will begin to develop. As the virtual strands manifest, their essences unfold.

These virtual DNA strands pair up, stack on top of each other on the completed double helix of the Youth and Vitality chromosomes.

Basic Stacking Order
Immortality
Creativity
E-Motion
Perfection
Communication
Physical
Cap top and bottom ends of chromosomes with telomere

k. Now begin to stack each pair of new virtual DNA strands on top of each other. *The healer may, at this time, be directed to celebrate aloud the purpose of each pair of DNA strands. It is important that Spirit direct you in the stacking order.* In most cases, the stacking order is presented below, but if Spirit directs you otherwise, then say whatever order comes through. You may also find that Spirit will add a few extra words to each stack as you speak:

- "I COMMAND THAT THE VIRTUAL DNA STRANDS OF COMMUNICATIONS BE STACKED ON TOP OF THE EXISTING DNA."
- "I COMMAND THAT THE VIRTUAL DNA STRANDS OF PERFECTION BE STACKED ON TOP OF THE EXISTING DNA."
- "I COMMAND THAT THE VIRTUAL DNA STRANDS OF E-MOTION BE STACKED ON TOP OF THE EXISTING DNA."

- "I COMMAND THAT THE VIRTUAL DNA STRANDS OF CREATIVITY BE STACKED ON TOP OF THE EXISTING DNA."
- "I COMMAND THAT THE VIRTUAL DNA STRANDS OF IMMORTALITY BE STACKED ON TOP OF THE EXISTING DNA."

l. The strands have their own color, with the top pair (immortality) having the gold/silver thread. Run this gold/silver thread—the *Thread of Everlasting Life*—through the stack, and wrap it throughout the entire stack of DNA.

NOTE: Depending on circumstances, as with children or pets, you can also use a shorter version: Command: "STACK THE VIRTUAL DNA STRANDS!"

m. The top of the new DNA connections must now be capped with telomere. Command: "I COMMAND THE ENDS OF THE CHROMOSOME STACKS BE CAPPED WITH TELOMERE." This caps the end of, and protects the chromosomes. Telomere helps the cells divide, replicate, and live longer. It looks translucent, like mother of pearl, and shines in shimmering rainbow colors. It fits over the top of the stack and coats the chromosomes. Capping with telomere is a vital step in the process. It has been proven that chromosomes with breaks in the telomere are subject to degeneration. A complete telomere wrapping strengthens the chromosomes and stops the aging process of the cell, thereby triggering lasting life in all cells of the body. By capping the chromosomes, we are pulling them into the electromagnetic field of the cell.

n. Finally command that the Youth and Vitality chromosomes be replicated throughout the body. You command: "I COMMAND THE YOUTH AND VITALITY CHROMOSOMES WITH THEIR VIRTUAL STRANDS TO BECOME WHOLE WORKING STRANDS WITHIN MY BODY. I NOW BRING THESE STRANDS INTO

BEING." At this point, you may see them become solid. Thus the DNA expansion has been set into motion. The client will notice a tingle or some other sensation, particularly in the head (crown chakra). This is confirmation that the DNA received your intent.

o. Upon completion of the rejuvenation, your higher-self will tell you that you have completed the task, either by a sense of calmness or silence, or you will get the message that it's time to separate.

Step 3: Ending the Session (common to all techniques)

a. Complete the healing by stating aloud: "GOD, THANK YOU FOR THIS OFFERING OF LOVE..."

b. Gratitude. We can never be too grateful for this universal love and service. By stating "GOD, THANK YOU FOR THIS OFFERING OF LOVE," we keep in balance the exchange of energy and remain in the oneness of our present state. It is okay to extend whatever additional words of gratitude that come to your mind.

c. "IT IS DONE" [repeated twice more, each time with more emphasis] invokes the magic of confirmation, acceptance, and realization. Now with the utmost passion say:.

"IT IS DONE ... IT IS DONE...IT IS DONE!"

d. Your higher-self will tell you when the healing is complete. Maybe a sense of calmness and silence will take over, or you'll get the message that it's time to separate. Leave the client's body through the top of the head [the crown chakra].

e. As you return to your body, wash (cleanse) yourself with pure white light. Imagine God's grace purifying you. Then perform an energy break from the client. This releases the client's energy from you, and pulls your own energy back into you. Polarity changes automatically each time you go

into the client's space, which is why the break is important otherwise you will feel "floaty" for a few minutes after separating from your client. Energy breaks are performed as follows:
- Rub your palms together vigorously for a moment.
- Extend one hand toward the client.
- Bring the other hand to your chest placing it over your heart chakra (center of your chest).
- Using your hand which is pointing towards the client, put it in a knife-like position, raise this hand to your forehead (Crown chakra), and make a slicing motion downward to your pelvic region (your base chakra), then raise your hand back up to your forehead.
- End by joining your hands in front of you as if in a prayer position.

This movement *breaks* the energy patterns between you and the client and *seals* your energy. It is absolutely necessary to perform a proper energy break, otherwise, you may permit the client's energy to move into your space. It is critical to cut and seal the energy after each session.

ENERGY STRENGTH TEST

Here is a fun way to see what happens with the energy and why it is so important to ensure you end the session with the energy break as described above.

1. Have the client stand with arms stretched toward you.
2. Have the client resist you as you push down on their arms. There should be lots of resistance.
3. Use your hand to "slice" through their energy field: start above their outstretched hands, go between their hands and end below them.

4. Have the client resist you (as in #2) and push down on their arms. This time, there will be little or no resistance.
5. With their arms stretched toward you, reverse the process of #3, i.e., start below their hands, go between their hands and end above them.
6. Repeat step one and two. There will be lots of resistance again.

A final word on the energy break with regards to the "Energy Strength Test" above. Make certain that after the test is completed that the person tested has regained his or her resistance in the arms. This signifies that the tested person has retained his/her power.

Children and Pets: Using the *short version* of DNA Activation™

Many people have asked me about children and pets. In most cases they are freer than we are and without a lot of baggage. Children and pets seem to have a natural tendency to magnetize me and the group whenever I've performed an ACTIVATION with a group. Children are easy to activate and quite patient. They seem to know what's happening. My seven year old daughter can describe the etheric light show and the angelic guides that are present. Children are open to the Activation. Pets also resonate with the energy fields in the room, are curious, and quite gentle. After the DNA ACTIVATION, both the children and pets become mellow and cuddly.

Performing a DNA ACTIVATION is quick for children and pets. You may want to use the *short version* below as the need for anything more lengthy seems not to matter.

In narrative form, the short version goes like this: In your mind or state out loud, make the following Commands:

"*GOD*, I COMMAND AN ACTIVATION OF THE YOUTH AND VITALITY CHROMOSOMES BE PERFORMED ON, _____ (person's full name), ON THIS DAY (the whole date), _____ AT THE TIME (morning, evening, or a.m. or p.m.). *GOD*, THANK YOU FOR THIS OFFERING OF LOVE ... IT IS DONE ... IT IS DONE ... IT IS DONE!"

Use the short version of the DNA ACTIVATION as intuitively you feel the need. It's all in the intent—that's the major requirement by Spirit.

Notes on the Activation Technique:

1. Please note that the images you see may be in shades of black and white. Over time, you will most likely be able to see colors and the actual organ, cell, and DNA.

2. Human cells have 46 chromosomes made up of the physical two-strand double-helix DNA plus the 10 rejuvenated virtual strands. These may appear fuzzy-looking, as if they've been smeared.

3. The DNA ACTIVATION™ Technique initiates many changes in the human body. Since it activates the Chronus and Vitality chromosomes which sets up new multiple-strands of DNA, new bodily structures will be formed. This causes an infusion of higher amounts of energy into the body while releasing obsolete and restrictive elements at the cellular level. Inform the client that he or she will most likely undergo surges of energy for several days or lethargic bouts and longer sleeping periods. Awareness of and preparation for this are important. Over a short period of time, the body will regulate itself, and the client will begin to sense positive physical, emotional, and psychological changes.

4. The time required for the DNA activation depends on the individual client. The healthier and more openhearted the

person, the higher the probability that the new virtual strands will orchestrate a new blueprint for the physical DNA restructuring and processing. Allow two to ten weeks for this process. Remember, it will take some time for the telomere to regenerate.

5. Regarding *stacking*:
   - It is important that the DNA multiple-strands stack in pairs upon each other in the order instinctive to the body.
   - Always stack in pairs.
   - Always command that telomere caps both ends of the chromosome.
   - NEVER CAP CANCEROUS CELLS.

6. Monitoring and charting: In most cases, people experience a spontaneous physical or motional healing from the activation. From that moment on, there will be numerous changes to experience over the next several months. We highly recommend that you journal and monitor the transitions. It's healthy to reflect on positive growth and not take them for granted. Use the "Monitoring Chart" in the back of the book to keep track of your changes and progress.

7. When ACTIVATING others, please make sure that they get the *DNA Healing Techniques* book. It serves as a reference and resource guide. ACTIVATIONS are real life events and changes are continuous. Having the book assists them to better understand the simplicity of the techniques and the magnitude of benefits that can be realized. Everyone ACTIVATED should have access to the book.

Conclusion

Now that DNA Activation is completed, the next step is to expand the newly activated DNA and chromosomes in every cell in the body (over ten trillion). This will be performed by using the following Expansion/Rejuvenation

Technique. This will complete the expansion of the newly regenerated multi-strand DNA in the remaining 44 chromosomes.

Technique #2: DNA Expansion and Rejuvenation

This technique activates, reconnects, and integrates the new virtual DNA strands with the original 44 chromosomes through all the cells in the body. This technique completes integration of the expansion process which began 2 to 10 weeks ago. The human body will come to reflect a uniqueness of its own: a higher vibrational Lightbody. In the process, any damaged or incomplete pieces of DNA will be repaired, and the function of the new virtual strands will begin the higher operational status of the physical body.

We need to change not only the Youth and Vitality chromosomes, but also the rest of the chromosomes. This is a direct result of the DNA expansion triggered by the initial activation.

Generally, human body cells reconstruct themselves every three months depending on the cell's function. At the heart of this process is the intricate task of reproducing the cell's DNA molecules. The two healthy chromosomes will cause the rest to eventually follow suit, and the energy of rejuvenation will spread to all parts of every cell. Several phone call follow-ups are highly recommended after the second activation.

The Expansion/Rejuvenation procedure requires utmost diligence in application. The second activation should be performed on a one-on-one basis. Do not perform the second activation with more than one person present at a time. You may perform a one-on-one second activation in a group setting as we do in our Certification classes.

Activation will become routine after several applications, so caution should be taken that the purest intent and love from the heart remain the healer's focus.

Guidelines for the Second Activation

How do we know if we are ready for a Second Activation? A much asked question that surrounds a quite ambiguous situation. Everyone is different and discernment plays an active role in determining if a person is ready for the Second Activation. There exists no set rules or formulas, but we do know a few great guidelines to go by. Before performing the Second Activation, whether to yourself or others, discern how many of the following apply for the person being activated"

- Was ample time allotted since the First Activation to demonstrate noticeable changes in physical, emotional, intellectual, and spiritual modalities? (Usually about two or more weeks.)
- Avoid doing the Second Activation if the person wants it because the just want to experience it. (This may be a sign of their ego and the "intent from the heart" not truly set.)
- Ask if the person has entered a peaceful state. (Hearing statements indicating a sense of peace is a key indicator to proceed.)
- Politely ask how does the person know if she or he is ready for the Second Activation. (Determine if the person is speaking from the heart.)
- Directly ask: "How do you feel about receiving the Second Activation?" (After that response, you intuitively sense to proceed or not.)
- Ask if the changes or transition since the First Activation have plateaued. (A positive response is a good indicator to proceed.)

These are only guidelines. Rarely do I get a "Do Not Proceed" message. Avoid being rigid in determining whether you should proceed with the Second Activation. It is a positive and beautiful event. If in doubt, probably best to proceed. Remember, it's Spirit's guidance that helps you determine whether or not to perform a Second Activation.

Step 1: Divine Orientation

Prepare yourself as you would begin a First Activation by performing the Divine Orientation.

Step 2: Specific DNA Expansion and Rejuvenation Technique

In your mind or state out loud, make the following command: "GOD, I COMMAND THE PERFECTION AND ACTIVATION OF ALL THE REMAINING CHROMOSOMES WITH THE MULTIPLE STRANDS OF DNA IN EVERY CELL OF THE BODY BE PERFORMED FOR_____ (person's full name), ON THIS DAY (the whole date), _____ IN THE TIME (early morning, evening, or am or pm) at _____ (location)."

a. In theta state, enter the client's body through the crown chakra and into the pineal gland. You only have to hold the thought in *theta* state for *two* seconds, then you can release it. Be very attentive to the inner *voice* of your highest consciousness.

b. Enter the great central cell. *See* the 10 virtual DNA strands stacked on top of the healed 2-strand double helix. The strands have their own color, with the top pair (immortality) having the gold/silver thread—the *Thread of Everlasting Life*—running through the stack, wrapping the entire stack of capped DNA. This provides you with a visual focus for the remaining work of this technique.

c. To complete the rejuvenation process, begin counting slowly from **1 to 46** allowing sufficient time for each chromosome to complete its activation.

This is a very critical process. The COUNTING is Spirit driven. You will be given each number or a series of numbers to say. If no number appears, then pause. Pauses between counts may occur to allow a specific healing to be processed. Wait until you see or feel the number as Spirit directs the call.

It is important to let your higher consciousness do the counting. Do not rush this process. Watch the activation spread to every chromosome. Sometimes you can actually see the process unfold. The person receiving the Second Activation will in most cases, be given information (or downloaded) beyond your recognition. That's normal. Sometimes you just count and when you open your eyes, you may see subtle or profound changes in the client.

Every effort should be taken not to place expectations on to the results of the activation. Enter each Activation without any pre-set limitations or expectations. In most cases, there will be emotional releases neither you nor your client are aware of. We usually have a box of facial tissues nearby.

Overall, the process is beautiful and we have consistently experienced pleasant and highly beneficial results. Treat these activations as a sacred process. You are doing God's work and in total service to the other person.

d. Once the chromosomes are activated and realigned, have the client say aloud: "I HAVE NOW ACTIVATED THE NEW CONNECTIONS IN MY DNA THROUGHOUT EVERY CELL IN MY BODY." Remember, by changing the DNA and chromosomes, we are changing the blueprint of the human body. We are actually reversing the effects of aging. The client will notice a tingle or some other sensation, particularly in the head (crown chakra). This is confirmation that the DNA received your intent.

f. You need to complete the connection between Father/Creator and Mother Earth to provide the electrical current that balances the male and female energies. We intensify communications with the Universal Energy and amplify the rejuvenation process at the cellular level, similar to a battery, now we hook up the connections. "I COMMAND THE RED/GOLD ENERGY OF MOTHER EARTH TO COME UP THROUGH MY FEET AND CONNECT ALL THE CELLS AND ENCOMPASS MY ENTIRE BODY."

g. " I COMMAND THE LIGHT GOLD UNIVERSAL CREATOR ENERGY TO COME DOWN INTO THE TOP OF MY HEAD (CROWN CHAKRA) TO CONNECT ALL THE CELLS AND ENCOMPASS MY ENTIRE BODY."

Once you have performed this, you have charged your body with electrical current. The healing process is so magnified that the body regenerates itself every three months instead of three years.

h. Upon completion of the rejuvenation, your higher-self will tell you that you have completed the task, either by a sense of calmness or silence, or you will get the message that it's time to separate.

Step 3: Ending the Session (common to all techniques)
[see Technique #1: Activation]

Notes:

1. Imagine these new DNA connections activated in every cell.

2. *ALWAYS bring in the DNA strands and chromosomes in pairs*. The connections in the strands must be done in **pairs**, and the virtual chromosomes must be activated in **pairs**.

Technique #3:
DNA Reading/Healing

The DNA Healing has many applications in reducing and eliminating acute and chronic disorders and impairments. Most individual seek the healing process because of disease, pain, and illnesses. However, experience tells us that the majority of issues relate to emotional crisis which in turn have caused the physical condition. A great amount of healing takes place, whereby many of the physical impairments begin to go into remission.

It is crucially important to understand that the practitioner is facilitating higher consciousness or Spirit's energy. We highly recommend that the practitioner avoid declarations promising cures or relief. The technique uses healing energies received from Spirit and is open to offer nutritional suggestions as well as using other healing techniques within the alternative medical and healthcare system.

We have also found that the more experienced the practitioner is as a healer, the more potential the healing facilitation. However, we firmly realize that the way of Spirit and the receptivity of the client are the governing factors when using these techniques.

This technique is to be used as many times as needed with a particular client for reading and healing. It can be used after a client has been activated. It is the healer's way of communicating at the cellular level of the body with regards to physical, emotional, and psychological matters.

Asking Permission

Remember, ask the client permision to do any form of healing work. This puts both the client and the practitioner in a state of balance and openness. Simply state: "_____" [client's name], "Do I have permision to do a healing with you at this time." We also recommend that you tape record the session and give a copy to the client for their reference. Healing sessions generally take 45 - 90 minutes.

Step 1: Divine Orientation

Prepare yourself as you would a First Activation by performing the Divine Orientation.

Step 2: The Reading/Healing (specific to this technique)

In your mind or state out loud, make the following command: "GOD, I COMMAND A READING/HEALING BE PERFORMED FOR _____ (person's full name), ON THIS DAY (the whole date), IN THE TIME (early morning, evening, or am or pm) at _____ [location]."

1. Picturing yourself as the etheric ball of light, enter the client's body through the head (crown chakra). As you begin to probe your client's body, you will be given information from both your higher-self and the DNA within the client's body.

2. *Imagine* what the inside of the client's head looks like. See the brain, the skull, the nasal cavities, everything inside the head. You may start by imagining a brain—this will be the client's brain. (Note: initially, you may only see variations of gray, but eventually, the actual coloring. It does not matter if you don't know the anatomy of the body accurately, you will see what you need to see.)

3. Most healers at this point can actually hear the *voice* of their higher-self coming through them. When this happens, you'll notice that you're not thinking, but rather allowing the information to come directly

through you. You should always remain conscious of what is taking place.
4. Move yourself through the client's body, lighting up the organs as you go. Again, you may imagine this in any way you wish, such as carrying a flashlight. If you are a ball of light, allow your *presence* to light the way. Any organ that doesn't light up has a problem.
5. Go through the body in any order, it doesn't matter—but the easiest is from head to toe. Usually, we start with the major organs, and go top to bottom, moving into muscles and the bloodstream, and finally into the skeletal system. You do this by willing yourself to the various parts, and by changing your focus in the body. You can view an entire organ, such as the heart, and then enter it to explore its inner parts, or change your focus and view the cells that make it up.
6. Remember, your cells talk to the client's cells, so ask the cells what is wrong. You will receive an answer. You may be shown the problem, you may hear a voice that explains what is wrong, or you may get a train of thought that lets you know. TRUST YOUR INTUITION. You are operating in the Light of God—you will know the answers.
7. With practice, you will be able to enter cells, and work with the components of cells. At times, immediately upon entering a body, you will be drawn to a certain area, and shown damage. *Talk to the client about the problem area, AND COMMAND A HEALING BE DONE ON IT.* Then you can explore the rest of the body at your leisure.
8. Healing is accomplished by *commanding* the part be healed, or by viewing the cells that make up the organ, or body part, and telling them they are whole, well, and functioning correctly. In the case of misbehaving cells, such as cancer cells growing in an organ, simply tell them they

need not grow, and command them to communicate with healthy cells. Remember, for healing on the physical or emotional body, *STATE WHAT IT IS THAT YOU WANT DONE* in the command.

9. Do not be afraid of doing the wrong thing. The work is overseen by the higher selves of you and the client, and harm will not be allowed to the client's body.

10. Sending light into affected areas is also useful, though not as effective as cellular healing. Surround and infuse the area with *green* light for healing, *pink* to relieve pain, and *white* for universal strength. Study what colors are good for certain ailments, and use them. You may simply command, "relief of all pain." This is useful for headaches, stomach aches, and the like.

Step 3: Ending the Session (common to all techniques)

[see Technique #1: Activation]

Notes on the Reading/Healing Technique:

1. If the client has ever been healthy, then you can command that the client be returned to his or her perfect blueprint.

2. (Optional - Spirit will guide you) For major healing, command: "GOD, I COMMAND THIS PERSON TO BE RETURNED TO THEIR ORIGINAL PERFECT BLUEPRINT," or if they were unhealthy at birth, "GOD, CREATE THE PERFECT BLUEPRINT FOR THIS PERSON AND RETURN THEM TO IT."

3. You need only hold the thought for healing for two seconds in a theta state. In meditation, using affirmations, telling an organ that it is healthy every day for a month, and believing it, you will have a

healthy organ. In theta, delivering the thought is delivering the results, and the organ, or cell, or body will listen. Even if you say to yourself, "I sure hope that worked," it will not effect the process in the least. In fact, the client doesn't even have to believe it works. You are bypassing the client's beliefs and speaking directly to the cells. However, if the client believes a body part is unhealthy, it won't take long to reprogram the subconscious into making that part unhealthy again. So, you may see dramatic results, only to find the same thing wrong in a week or two.

4. If you are working on any major illness or disease, you may be directed to use the Activation or Expansion/Rejuvenation techniques before proceeding in order to enhance the communication of the healing process of the cells. Listen carefully to the *voice* within, even if the information is presented slowly.

5. Regarding clients with *cancer:* in the presence of toxins, viruses, mutagenic substances, or radiation, a "rogue cell" may result. This is usually detected and destroyed by the immune system, but if not, the rogue cell may proliferate to form a tumor at the expense of the healthy cells, i.e., cancer. Cancer cells produce the enzyme telomerase, which restores the telomere. If your client has been activated by the Activation and Expansion/Rejuvenation techniques and has a bout with cancer, concentrate future healings directly into the cancer cells commanding them to neither produce telomerase nor replicate.

6. Leaving through the client's crown chakra also puts you in position for psychic readings. You can plug into the client's subconscious mind or talk to the higher-self. Put your consciousness about three feet above the client, and simply command a reading. The

information you receive will be as accurate as the work you've done in the body, and you'll receive an important message for the client.
7. This technique can be repeated for specific healings as often as is needed by a particular client or yourself.
8. DO NOT CAP CANCEROUS CELLS!
9. Regarding Viruses
   a. Every virus has a tone. When performing the Reading/Healing Technique and a virus is present, ask for the tone to come through that *disempowers* that virus.
   b. As a virus becomes stronger it may be necessary to use more than one tone because a single tone may not work anymore.
   c. Chromosomal diseases are caused by defective DNA cells that build walls around them causing them not to communicate.
   d. When you destroy tumors and unhealthy cells, command the debris to be turned into healthy cells again or leave the body immediately.
10. Cancer Healing Treatment Approach
    a. Approach client with unconditional love.
    b. Keep calm (avoid being nervous and do not panic). Go into *theta* to facilitate the healing.
    c. Cancer cells are caused by viruses which mutate cells.
    d. Go meet the cancer cells. (Don't treat them as bad things, but somewhat ignorant and they need to communicate with other cells).
    e. Once you know the cells, talk to them.

f. Go in and command cancer cells to break down the walls of virus/cancer cell so that they can communicate with healthy cells again.
g. Check under lobes under arms and liver (lymphatic system).
h. Command tumors to dissolve.
i. Use color therapy (green & red or whatever spirit directs you to use).
j. Ask Creator for the tone that will destroy the virus.
k. Listen to tones and repeat these tones.
i. As soon as cancer cells communicate, they begin to die.
m. You want to turn cancer cells into normal cells, not destroy them.
n. See client as often as necessary (2-3 times per week).
o. Document the progress.

IMPORTANT: with these simple steps, you'll be able to bring relief to people with dis-ease in their bodies. Do not be afraid to change specifics to suit you and the situation—everyone has his or her own style. *You can also use this technique in conjunction with other techniques you may already be familiar with.* As with everything, practice makes perfect, and the more you do this, the better you'll facilitate the healing process. Always approach healing with *love* and *good intent* and *from the heart*, and you'll be amazed with the results.

Part 4: Transitioning after DNA ACTIVATION

"The DNA Activations and healings accelerate Spirit's relentless drive to lovingly resolve issues and purify the whole being."

In this section, we will discuss the transitioning that occurs after a person has received both DNA ACTIVATIONS™. Spontaneous healings can occur at the time of Activation, most of which relate to acute physical and emotional disorders. It appears that the First DNA ACTIVATION works to get the physical body realigned toward a healthier state of existence. The Second DNA Activation works to realign the psychological improvements of the individual.

Most Activated people have reported consistent changes, some more profound than the others, but most noticeably, changes are occurring and we are able to see very positive trends. Though most reported changes appear related to physical conditions, emotional release, intellectual clarity, relationship improvements stand significantly at the forefront of those in transition.

With regard to the *Reading/Healing Technique*, we are gathering more information based on hard facts that support improvements and healings with clients who have cancer, fibromyalgia, diabetes 2, bone fractures, bodily pain, weight loss, addiction, and other ailments. In all cases, DNA Practitioners work under strict ethical codes of conduct and do not interfer with current medical or psychological treatments of the clients.

Reported Results, Symptoms, and Side-effects

By expanding the DNA, we are changing the make-up of the human body. We are actually reversing the effects of aging. There is increased activity in immune systems, bones heal faster than normal, diseases clear up. Cell regeneration speeds up, and degeneration slows. While it's too early to tell, life may even be prolonged.

There are some *side-effects* that you and your clients should be aware of. Existing problems may flare up. Anything wrong with the body is brought up to the surface and your clients may go through a period of increased dis-ease. Flu-like symptoms and emotional release are the most common. These reactions are similar experiences to those of cleansing diets or fasting.

The client may develop a rash, usually around the thyroid region which may last for a few days and then go away for good. There may be a few days of increased energy, to the point of sleepless nights. This is sometimes followed by a three day period of decreased energy or lethargy, but when it goes away, the client will feel more energized than ever before. Also, the increase and decrease of energy may reverse, with lethargy preceding too much energy. Within several weeks, however, all symptoms will disappear, and life will return to normal. Clients may also notice an increase in psychic ability.

To date, some of the physical benefits reported by clients include:

- wrinkles fade
- body detoxifies
- weight stabilizes
- feel younger
- releasing metals
- less stress and worry
- eyesight improves
- patience improves
- muscle renewal
- nails grow faster
- skin tightens
- hair grows and thickens
- stretch marks fade
- energy level increases
- renewed sense of competence
- sense of security
- varicose veins diminish
- improved digestion
- sharper memory
- "You're different" remarks

Some of the more subtle psychological and physiological results that can be reported are grouped into the categories presented below and are further discussed in the *Discover Your Destiny* Training Programs. Immediate changes may include:
- heightened perception
- precise use of language
- increased desire for pure water
- desire for better nutrition
- happy, happy, happy
- enhanced discernment
- increased telepathy
- lucid dreaming
- less need for food
- increased self trust

The DNA Activations truly bring forth a clearer orientation to life and personal destiny. It accelerates self-mastery and opens the door to greater possibilities of who you really are, that is knowing your "Higher-Self." In that process, the quest to find and better express yourself is heightened. Below are a few changes that have been reported:
- passing on the "other side" phenomena
- what does life feel like?
- sensations of bliss
- quicker manifestation
- not allowing old stuff to register
- replacing "seeking" with "finding," then "resolving"
- experiencing "the Crucifixion" (reliving sorrow, trusting death, and allowing God's hand to take you into immortality)

Improving the state of all our relationships takes a priority for most people. We are on this planet to experience emotions, create, and share. Sharing in relationships can be formidable. With the DNA Activations, a keener resolve for relationships exist. Below are a few changes that a person may experience while in the transition:
- new concepts and awareness to self and others
- allowing new feelings to be stated and shared

- letting go of old stuff
- seeing relationships more clearly
- feeling as a whole being (oneness)
- finding deeper love with others
- experiencing rapture with or without sexual orgasm
- attracting "soul mates and soul family" members

Self-discovery is a key to our freedom. Looking within to find the road-blocks of life is a courageous challenge before each of us. Everyone of us has these so called "Self-Truths" (fears, doubts, traumas, hurts, memories, and emotions) which are obstacles that interfere with our true expression of "self." The DNA Activations delve directly into these "Self-Truths" and strive to eliminate them by:
- the automatic capacity to confront the self-truths
- finding answers that have meaning and clarity in life
- not allowing uncertainties to dictate the accuracy and truth in situations
- finding the feelings, intuition, and communications that keep us focused and in the moment.

Testimonials

Proof. Everyone wants proof. "Demonstrate to me that all this works." Yet we constantly seek and hope that something will be available to cure our ailments. The mysterious drugs are trust and faith. When a few take the leap forward in life and show by example, the rest will follow, and following is dependency, and dependency is the slow way. The DNA ACTIVATIONS are but one way to break the barrier of conformity and experience one of the greatest gifts of our time.

In the back section of this book are but a few of the many testimonials received as a result of the DNA Healing Project and we expect millions more.

Meditation

Meditation not only provides the doorway to relaxation, it serves as the communication link between your soul and your mind. In meditation, a person can receive valuable input on how to cope with life's issues and insights on what may be forthcoming. Over the course of the DNA Healing Project, I have found that the DNA ACTIVATION Technique also serves as a great way to launch yourself into a meditation which I easily titled: The DNA Meditation.

It's quite simple and uses the *Divine Orientation* process. If you desire to resolve an issue or bring in supportive input, use one of the virtual DNA strands that closely relates to your situation to help get a better handle on the topic. For instance, if I have an emotional dilemma, I would call in the DNA Strands of E-Motion. If I was depressed and felt alone, I would call in the DNA strands of Immortality.

A dilemma example: "God, I command that the DNA strands of E-Motion assist me in dealing with my dilemma ____(?)____. Thank you God, for this offering of love. It is done. It is done. It is done!"

A depression example: "God, I command that the DNA strands of Immortality assist me in dealing with my depression. Thank you God, for this offering of love. It is done. It is done. It is done!"

An abundance issue example: "God, I command that the DNA strands of Creativity assist me in manifesting abundance. Thank you God, for this offering of love. It is done. It is done. It is done!"

Be creative and let meditation serve you. It's a valuable tool and the more it's repeated, the more confident you become about its function.

Nutrition

The DNA ACTIVATIONS automatically set forth the process requiring you to improve your health. Reports and testimonial affirm this. Within a few days, people who have been activated drink more water than before their activations. They find resolve in knowing that proper nutrition becomes a priority. Transitioning after DNA Activations drives the individual to eat better and focus on keeping the human machine a sacred place for the soul to dwell. We have found that the body desires more proteins, amino-acids, minerals, omega fats, and less carbohydrates, unless very active.

Knowledge and adaptation of proper nutritional practices will definitely augment a healthier body and mind. Eating a good balanced diet is essential. Two great best-selling books, *THE ZONE* by Barry Sears, Ph.D., and *Eat Right For Your Type (blood type)* by Dr. Peter J. D'Adamo clearly state the necessity of proper food intake and balance. In brief, Sears suggests the 40-30-30 rule, respectively carbohydrates-protein-fats, best serve the body's nutritional requirements. D'Adamo suggest the your blood type is crucial in determining what nutritional program best serves your body. Both are viable and work. I highly recommend you seek the advice of a respected nutrition and begin reading some good books on the subject. What is "soul food" anyway?

Exercise

Yes, exercise is essential. Stretching is essential. Breathing is essential. Your body requires exercise to facilitate the electro-chemical process. Exercise indirectly acts as a regulator. Without minimal exercise, the body looses its ability to regulate proper balance between the food and cellular growth. Every effort must be made to let your body know that you care and respect it. The DNA Activations serve this process and you will notice less resistance to exercise within a week or two.

Epilogue

We believe that these techniques will help people live longer and healthier lives. Applying these techniques will aid transformation into the next level of consciousness and possibly higher dimensions of *life*. Learn this, and do this. With *love in your heart*, change the DNA in yourself and everyone you meet, raise their vibrations, and help heal mankind and the planet.

This second edition of *DNA Healing Techniques* includes new information simplifying the presentation of the three DNA Healing Techniques. The modifications within were made from the experienced gained and feedback received from our Certified Practitioners and clients. Every effort has been made to keep the techniques simple yet effective. Supportive information was added offering the reader and new practitioners necessary detail to better explain and deliver the techniques. We will continue to expand into more editions as new techniques for specific ailments are identified and testimonials received.

We encourage you to report to us what effects these techniques have on you and your clients. We have prepared a DNA Monitoring Chart to record physical and psychological changes observed. A modified version is included in the back section of this book for your daily use.

It is vitally important that you document every activity with each client. A *Client Case Report* has been prepared and is available *free* of charge. Just call us. We will assemble case reports by category and will publish extracts from them to share the wealth of healing information around the world. To receive the latest information about the DNA Healing Project, please subscribe to our new "InSights" Journal. Subscription information appears in the back of this book.

The Spanish translation has been completed. When appropriate funding becomes available, the Spanish Edition will be released. Negotiations are underway with German, Japanese, and Hebrew translators. As resources become available, these and other translations will be published.

We can always use your help and support. As you discover how well these DNA techniques work for you, help us get this message out to others. Those in the medical and scientific professions who see the benefits of these techniques are encouraged to contribute their wisdom and participation in our programs. We could truly use your support.

Finally, please join our Foundation and become a member. That financial contribution would be truly welcomed, and it is the easiest way for you to show your gratitude, to stay involved, and to keep up-to-date with all that is happening with these simple but extraordinary healing techniques.

— Blessings from Vianna McDaniel, Dr. Todd Ovokaitys, and Robert Gerard.

Certified Training Programs

It is important that these DNA Healing Techniques are spread around the world. Once you've become adept, you may begin to teach others. You will find it is simple and comes easily to most people. We are offering low cost Certified Training programs at the Foundation headquarters in Livermore, California, soon regionally, and eventually, nationally.

The intent of certification is two-fold: to keep the techniques consistent and to gather your feedback for our research. As we receive and document Case Reports from around the world, we will be able to pin-point healing procedures for specific abnormalities. This will be accomplished with the highest regard for the individual and with utmost integrity.

The certification process has been designed to support the Foundation's DNA Project Mission: "to activate healers, teach the DNA Healing Techniques worldwide, and promote ongoing research into new ways of using the techniques." By certifying DNA healers, the Oughten House Foundation can assure the integrity of the DNA Healing Techniques, protect their simplicity, keep them authentic, prevent fraud and commercialism or misuse, and generate marginal revenue to keep the worldwide expansion progressive.

Three categories of Certification have been established to better facilitate the public, the Certified DNA Practitioner, and the Foundation.

- Practitioner in DNA ACTIVATION™ Techniques
- Practitioner in DNA Healing Techniques
- Senior Practitioner in DNA Healing Techniques

Each category stands on its own merits, distinctive of the others. Each based on healing performance competencies and the business relationship commitment to the Foundation. Progression to more advance certification is sequential.

Each certified DNA Practitioner serves as a representative of the Foundation. We will do our best to make you more visible in your community and strengthen your compensation as the Foundation progresses toward its goals with your support and affiliation. This interchange of values, recognition, and compensation are basic requirements for both parties. The certification tuition and renewal fees established are part of the financial exchange designed to keep the Foundation proactive and, in return, keep the Certified Practitioners pro-active as well.

To receive more information about the Certification Programs, course schedules and tuitions, please contact the Foundation.

Discover Your Destiny Programs

Training Programs & Overview

Our primary purpose on this planet is self-mastery: knowing yourself and your destiny. As you embark on this journey, your primary objective is learning how to live in the moment. Your past and your future are truly revealed to you. Freedom of expression and expressing Spirit are the results. Living life in this mode increases your perception of your inner and outer worlds. You become empowered, enlightened, and the embodiment of Divine Consciousness (the Christ Consciousness) envelops you. You, the Master, are one with GOD.

The purpose of the *Discover Your Destiny* (DYD) Educational Programs is to develop the "Master" that you are. This process begins with your commitment to open your heart. To become vulnerable and transparent to yourself and others, and allow the depths of your soul to open up and reveal its secrets. As this self-mastery process evolves, uncompleted karma will surface; desires and thoughts which contain judgments of self and others will return to be experienced or eliminated, and you must confront that which you have consciously or subconsciously denied.

There are many doors on the path back to God, each carefully designed by your soul as part of the Divine Plan, stepping stones that reveal self-truths—positive or negative—and provide opportunities to face these self-truths and to discern how to confront them or honor them. A journey of revelations, if you will.

As you will discover, the intent of the Foundation's *Discover Your Destiny* educational programs focuses on developing your self-mastery. The healthier you are in mind, body, emotion, and spirit, the healthier will be mankind and the planet.

Each of the lectures, seminars, and workshops is designed to empower and enlighten you. Each program has been specifically designed to provide tools and techniques that encourage the individual's self-mastery, the discovery of personal destiny, and foster the evolution of the Lightbody.

PARTIAL LISTING OF SEMINAR AND WORKSHOP TITLES
DNA#101 Basic Overview of the DNA Techniques
DYD#103 Confronting Self-truths/Basic
DNA#105 Identifying the New You: Basic Transitioning

DYD#320 Discover Your Destiny / Intro Seminar
DYD#330 Discover Your Destiny / Advanced Seminar

DYD#501 Speaking From Your Heart Lecture
DYD#505 Accurate Communications/Confrontation Intro Lecture
DYD#510 Confrontation Skills Training Intro Workshop* (1-day)
DYD#525 Confrontation Skills Training Public* (weekend)
DYD#550 Confrontation Skills Training Commercial Programs*
   * Courses related to communication and confrontations are based on the book *Handling Verbal Confrontation: Take the Fear Out of Facing Others* by Robert V. Gerard, ISBN 1-880666-05-7 $14.95

Tuition fees will be kept as low as possible. For locations, dates, and details of the *Discover Your Destiny* Programs, please contact the Oughten House Foundation. Your tax-deductible donations to help us research and network these DNA Healing Techniques are always welcomed and appreciated.

Certified Instructors and Independent Contractors

Recruiting individuals to see that the DNA and DYD Educational Programs expand worldwide will be a strategic aspect of the business. The people, the healings performed, and the education delivered perpetuate the organization and business behind the mission. The recruitment of heart-driven people, healers, and educators dedicated to the mission is key. Compensation schedules and fee structures are based on program and certification criteria.

Independent contractors interested in affiliation with the Foundation, should furnish a letter of intent addressing the specific topics or programs you are interested in, a resume if available, some pertinent background information, and three references to: Robert Gerard, Oughten House Foundation, Inc., PO Box 3134, Livermore, California, 94551-3134 USA (phone 925-447-2372, fax: 925-798-5848), or E-mail the same to: dawne@value.net, or Robs1World@aol.com."

"InSights" Journal

Our new publication "InSights" is an interactive journal that offers readers and practitioners the latest updates as to the events and healings surrounding the DNA Healing Project, its mission, and growth. It is a forum for individuals who desire to place discrete advertisements, feature articles, the latest information on using the techniques, and will have a *Contact List* of the Certified DNA Practitioners worldwide. Annual subscription is $25 for this by-monthly tabloid.

Suggested Readings

Ageless Body, Timeless Mind, Deepak Chopra
Brain States, Tom Kenyon, M.A.
Eat Right for Your Type, Dr. Peter J. D'Adamo
Everything is Now and *Creative Conflict*, Dr. Christopher Hill
From Genes to Cells, Stephen R. Bolsover, et al
God, Cosmos, and Man, Wayne Fields, Ph.D.
Handling Verbal Confrontation, Robert Gerard
Hands of Light, Barbara Ann Brennan
Healthy Living: A Nutritional Guide, Linda Rector Page, Ph.D.
Human Genome Project, Elizabeth L. Marshall
Kirael: The Great Shift, Fred Sterling
New Bodies, New Cells, New Life, Virginia Essene
Spontaneous Healing, Andrew Weil, M.D.
Solarian Legacy, Paul Von Ward
The Human Body: An Illustrated Guide to Its Structure, Function, and Disorders, Editor-in-Chief Charles Clayman, M.D.
The Kryon Writings, Lee Carroll
The Awakening—Eternal Youth, Vibrant Health, Radiant Beauty, Patricia Diane Cota-Robles
Walk Between Worlds, Greg Braden

Book Discounts

The price of this book is $8.95, plus S&H of $2.50 for the first book, and 50¢ for each additional book. Order 20 or more copies for a 20% discount. Donations to Oughten House Foundation, Inc. are tax-deductible and will help fund further research, newsletters, and future correspondences.

Don't forget! Send for your free Case Study Report Sheet. (Please include a SASE #10 envelope.)

Audio Tapes Available

Experience aspects of the DNA Healing Project and DNA Activations. Each tape carefully produced with you in mind. They are informative, straightforward, and realistic. Activation tapes are encoded with "high energy from Spirit" to bring you an ultimate experience. Listeners have positively commented on their results obtained form the Activation Audio tapes.

The DNA ACTIVATION™ Tape

Includes the DNA meditation, the Soul Bubble Visualization exercise, the First DNA Activation, and commentary on the DNA Healing Techniques. This audio tape contains the energy to thoroughly experience the Activation. — 60 minutes, $20.00

The Second DNA ACTIVATION Tape

Includes the Pulsar Visualization Theta exercise, the Second DNA Activation, and commentary on the transitioning resulting from the second activation. This audio tape contains the energy to thoroughly experience the Activation.
— 45 minutes, $18.00

The DNA Healing Project Overview Tape

Includes insights into the mission plan and each technique.
— 15 minutes, free

The DNA ACTIVATION™ Book & Tape Set

Includes the DNA Healing Book ($8.95), the First DNA ACTIVATION tape ($20) and the Second DNA Activation tape ($18), plus additional information. This book/audio tape set valued at $46.95 is discounted to $40.00 (Save $6.95).

The DNA ACTIVATION™ Tape Set

Includes the First DNA ACTIVATION tape ($20) and the Second DNA Activation tape ($18), plus additional information. This 2-tape audio tape set valued at $38 is discounted to $33.00 (Save $5).

Shipping & Handling: $2.50 for the first tape, $0.50 each additional tape or tapes within sets.

About the Authors

Robert Gerard, publisher of Oughten House Publcations, a leading spiritual publishing house has written *Lady from Atlantis*, *The Corporate Mule*, and *Handling Verbal Confrontaion*. Robert, in the role of CEO, visionary, lecturer, and healer, serves Oughten House Foundation, Inc., which is devoted to education and worldwide networking of the DNA Healing Techniques and the "Discover Your Destiny" educational programs. He may be reached by contacting Oughten House.

Dr. Todd Ovokaitys holds a medical degree from John Hopkins University. He has pioneered DNA laser research and has been quite active in the fight against AIDs. Dr. Ovokaitys can be reached at Gematria Products, Inc., 2075 H. Corte Del Nogal, Carlsbad, CA 92009. Phone: (760) 931-8563, fax: (760) 931-8492, email: richard@gematria.com.

Vianna McDaniel, the pioneer of the DNA Healing Techniques, is part of Nature's Path, a group of three psychic healers, devoted to bettering mankind. Vianna, a Minister of Religious Science, a nutritionist, and Rieki Master offers training, classes and workshops, and may be contacted at 1920 E. 17th St. Suite 115, Idaho Falls, ID 83402. Call (208) 524-0808, 9 to 5, Monday through Friday.

To find out about the DNA Healing Project, training, classes and workshops, contact Oughten House Foundation, Inc., P.O. Box 3134, Livermore, California 94551-3134 USA, phone (925) 447-2372, fax: (925) 798-5848, E-mail: dawne@value.net, or visit our Internet Web Site: www.oughtenhouse.com.

DNA Monitoring Chart

Positive alterations in mind, body, emotions, and spiritual knowingness begin to occur immediately after your DNA Activation. This will be gradual for about five weeks or upon receiving the final DNA Expansion/Rejuvenation Technique. These alterations will then be heightened more as your multi-strand DNA prepares you for living in the realms of higher consciousness.

It is important to journal and monitor your progress. Below are four areas that deserve attention. Use this chart to monitor your progress over the next four weeks. Rate "1" low to "5" very high in the appropriate box to indicate your progress. Place and "X" in the cell if no changes were perceived. Add more categories as needed.

On the back, please give us in your own words, what changes you have experienced and how, emotionally you have dealt with them. We appreciate receiving a copy of this chart to aid us with our research.

Don't forget! Send for your free Case Study Report Sheet. We will also send additional information that is available. (Please include a SASE #10 envelope.)

DNA Monitoring Chart

Transitioning Experienced	Weeks after Activation			
	1	2	3	4
Dates:				
Becoming aware of:				
Increased desire for pure water intake				
Enhanced self-awareness				
Precise use of language (conscious languaging)				
Better nutrition/less food				
Enhanced discernment				
Heightened perceptions				
Increased telepathy and clairvoyance				
The "Crossing Over" phenomena:				
Easier & quicker connection with Spirit				
Allowing God's hand to take you into immortality (Experiencing the relief of sorrow and death)				
Replacing "seeking" with "finding"				
More centered in the "present"				
Peace and calmness within				
"Old" truths no longer work				
Frequent lucid dreaming				
Quicker manifestations				
Sensations of bliss				
Keener resolve for relationships:				
New concepts and awareness to self and others				
Allowing and sharing new feelings				
Seeing relationships more clearly				
Finding truer love with others				
Welcoming new insights				
Experiencing rapture				
Letting go of old stuff				
Confronting the Self-Truths in situations:				
Finding answers with meaning and clarity				
Accuracy and truth dictating situations				
Focusing on finding resolutions				

Technique #1: DNA ACTIVATION Reference Guide

Technique #1: Activation of the Youth and Vitality Chromosomes (also known as the DNA ACTIVATION™)

Two archetypal chromosomes are primarily responsible for activating the 12-Strand DNA to its full potential. They are The "Youth" and the "Vitality" chromosomes. It's best for you and the client to sit facing each other. Place your palms under the client's palms. This creates an electrical circuit, and allows your cells to begin talking.

1. Raise your consciousness

   - Become centered in yourself, take a few deep belly breaths, pull all of your energy inward, focusing it in your heart. Balance your energy. Feel at peace.

   - Let your consciousness rise from your heart through the top of your head, about six feet above your head.

   - CALL IN GOD, [or whatever term you prefer to address your deity]. Once you do this, you go into a *theta* state.

2. Make the following Command, "*GOD*, I COMMAND AN ACTIVATION OF THE YOUTH AND VITALITY CHROMOSOMES BE PERFORMED ON, _____ (person's full name), ON THIS DAY (the whole date), _____ AT THE TIME (morning, evening, or a.m. or p.m.) at _____ (location)".

   The preceding wording triggers an activation for the client. You have to hold the thought in the *theta* state for only TWO seconds. In this *theta* state, enter the client's body through the top of their head and into the pineal gland. Be very attentive to your inner *voice*.

3. Command: "SHOW ME THE GREAT CENTRAL CELL."

4. Command: "SHOW ME THE CHROMOSOMES."

5. Command: "SHOW ME THE YOUTH AND VITALITY CHROMOSOMES."

6. Command: "SHOW ME THEIR DNA."

7. Have the client repeat: "I COMMAND ACTIVATION OF THE YOUTH AND VITALITY CHROMOSOMES IN ME NOW!"

8. Now begin to stack each pair of new virtual DNA strands on top of each other:

   - "I COMMAND THAT THE VIRTUAL DNA STRANDS OF COMMUNICATIONS BE STACKED ON TOP OF THE EXISTING DNA."
   - "I COMMAND THAT THE VIRTUAL DNA STRANDS OF PERFECTION BE STACKED ON TOP OF ALL MY EXISTING DNA."
   - "I COMMAND THAT THE VIRTUAL DNA STRANDS OF E-MOTION BE STACKED ON TOP OF ALL MY EXISTING DNA."
   - "I COMMAND THAT THE VIRTUAL DNA STRANDS OF CREATIVITY BE STACKED ON TOP OF ALL MY EXISTING DNA."
   - "I COMMAND THAT THE VIRTUAL DNA STRANDS OF IMMORTALITY BE STACKED ON TOP OF THE EXISTING DNA." Run their gold/silver thread—the *Thread of Everlasting Life*—through the stack, and wrap it throughout the entire stack of DNA.

9. Command: "I COMMAND THE ENDS OF THE CHROMOSOMES STACK BE CAPPED WITH TELOMERE."

10. Command: "I COMMAND THE YOUTH AND VITALITY CHROMOSOMES WITH THEIR VIRTUAL STRANDS TO BECOME WHOLE WORKING STRANDS WITHIN MY BODY. I NOW BRING THESE STRANDS INTO BEING."

11. Show gratitude. "GOD, THANK YOU FOR THIS OFFERING OF LOVE." and with the utmost passion say: "IT IS DONE ... IT IS DONE ... IT IS DONE!"

12. Perform an energy break (cleanse) from the client.

Technique #2: DNA Expansion and Rejuvenation Technique Reference Guide

1. Raise your consciousness
   * Become centered in yourself, take a few deep belly breaths, pull all of your energy inward, focusing it in your heart. Balance your energy. Feel at peace.
   * Let your consciousness rise from your heart through the top of your head, about six feet above your head.
   * CALL IN GOD, [or whatever term you prefer to address your deity]. Once you do this, you go into a *theta* state.
2. In your mind or state out loud, make the following command: "GOD, I COMMAND THE PERFECTION AND ACTIVATION OF ALL THE REMAINING CHROMOSOMES WITH THE MULTIPLE STRANDS OF DNA IN EVERY CELL OF THE BODY BE PERFORMED FOR_____ (person's full name), ON THIS DAY (the whole date), _____ IN THE TIME (early morning, evening, or am or pm) AT _____ (location)."
3. Enter the great central cell and see the 10 virtual DNA strands stacked on top of the healed 2-strand double helix and capped with telomere.
4. To complete the rejuvenation process, begin counting slowly from **1 to 46** allowing sufficient time for each chromosome to complete its activation.
5. Have the client say aloud: "I HAVE NOW ACTIVATED THE NEW CONNECTIONS IN MY DNA THROUGHOUT EVERY CELL IN MY BODY."
6. "I COMMAND THE RED/GOLD ENERGY OF MOTHER EARTH TO COME UP THROUGH MY FEET AND CONNECT ALL THE CELLS AND ENCOMPASS MY ENTIRE BODY."
7. "I COMMAND THE LIGHT GOLD UNIVERSAL CREATOR ENERGY TO COME DOWN INTO THE TOP OF MY HEAD (CROWN CHAKRA) TO CONNECT ALL THE CELLS AND ENCOMPASS MY ENTIRE BODY."
8. Show gratitude. "GOD, THANK YOU FOR THIS OFFERING OF LOVE." and with the utmost passion say: "IT IS DONE ... IT IS DONE ... IT IS DONE!"
9. Perform an energy break (cleanse) from the client.

Testimonials

"Many changes occurred: wrinkles faded, skin tightened, hair grew fast, weight stabilized, stretch marks are beginning to fade. I have lots and lots of energy, patience and am 'happy, happy, happy'. My memory is sharper and I don't stress and worry about daily events."
— Katie W., Hospice Social Worker, Idaho

"I am 74 years young and since my activation, I am no longer in despair, my skin is tightening and the wrinkles are disappearing. I am gardening again and sleeping all night." — Judy H., Retired, California

"I have had Fibromyalgia for 17 years and since my activations and healing work with these techniques, I don't feel depressed or tired like before. I have a lot of positive feelings and lots of energy. I sleep better and my whole outlook has changed. My eyes have changed from dark hazel to blue!"
— Peggy H., Homemaker, Idaho

"Painful childhood memories came up and were released. It was preverbal and I believe it was in the birthing process. I have had tremendous emotional release without emotional attachment. My dreams are extremely vivid and the meditations are powerful. The light in my meditations is brighter and I feel the colors are communicating to every cell, atom, and organ in my body. I require less sleep and my mind is clear."
— Sheila B., Promoter, Hawaii

"Everything that does not pertain to my highest good and purpose has lovingly gone away. Emotional events that have prevented progress on my mission have been revealed and removed, severing forever the ties that have bound me. My photographic memory is being restored after having been shattered by 'life' events. My purpose, mission, and destiny have been clearly revealed to me. The love and trust of self is total. I am at peace, finally." — Aerial F., Financial Director, California

"I am now on day 17 of approximately two hours of sleep each night. At first I saw it as out-of-the-norm and somewhat annoying, but when I realized that I had no fatigue and even plenty of energy for the gym, I finally surrendered to the process. Since then, I have tapped into a stream of consciousness that is entirely spirit-based; and I've come to realize it was part of me all along. I feel clear and at peace. I feel happy and full of energy. I feel as though the child-like essence of innocent wonder at the life-process has returned, magnified, and continues to expand. Working consciously to witness this Self-Mastery process is an amazing, exciting and incredibly rewarding experience. The DNA Activation was an integral part of my awakening." — Catherine B., Publicist, Arizona

"Something is happening. Something BIG, I know it! I have Fibromyalgia and don't remember ever being without pain. I actually have days when I have no pain, and when the pain is there, I am disassociated from it. I have so much energy now and I can physically do things I have been unable to do for many years. My mission has been shown and I feel great!" — Larry P., Reborn Activist, California

"My relationship with my wife has changed for the better. I am able to talk to her without anger." — Dr. Fred R., Pathologist, California

"I have had a tremendous fear when someone else drives. Since my activation, however, I was able to talk in a rational and reasonable way to my husband about my fear when he was driving, and we were actually able to communicate and laugh about it. This has never happened before." — Lois S., Homemaker, California

"I have been able to release and let go of many things. My mind is clear and my emotions are elevated to love of humanity. I have "crossed over to the other side" and many visions have been shown to me." — Bob G., Advisor, California

I am better able to speak my feelings and thoughts with clarity and love. I have experienced better communication with precise use of language. Awareness and trust in myself has increased. I live in the moment while still aware of my responsibilities. I don't sweat the small stuff anymore." — Lorena K., Business Owner, California

I have a higher sense of accomplishment, less stress and more energy. I have love and acceptance for myself and others." — Nina B., Sr. Medical Technician, California

My awareness, telepathy, intuition, and connection with spirit has improved greatly since doing the DNA work. I have improved nutrition. My sleep patterns and lucid dreams improved immediately." — Valerie D., Marketing Director, California

Fund Raising and Donations

The DNA Healing Project and Mission is REAL. Over a thousand people have been activated with invariably positive results and well-being. There's no denying that something wonderful and beneficial is happening. We can use all the support possible. We have books to translate and print, people to heal, and people to teach how to heal. We make no claim to fame, just simple love through simple techniques. Your support is appreciated.

Please help! Any donation or contribution is appreciated. Our staff of volunteers is overworked. Our mission is strong. Our service heart-driven. Please help us network these fine results oriented healing techniques around the world.

Oughten House Foundation, Inc. operates as a non-profit, tax exempt 501(c)(3) educational organization. Most donations are tax-deductible. Major credit cards accepted. Receipts generated. Send your donations to the Oughten House Foundation address below.

Foundation memberships are available. Individual adult memberships are $30. Membership fees vary. For benefits, enrollment information, subscriptions to the "InSights" interactive journal, and membership enrollment kit and free gift, please write or call.

—Blessings.

Oughten House Foundation, Inc.
PO Box 3134, Livermore, CA, 94551-3134,
Phone: (925) 447-2372, fax: (925) 798-5848
E-mail: Robs1World@aol.com
E-mail: dawne@value.net

Notes

Notes

Notes

Notes

Notes

Notes

Notes

Notes